This book is a presentation of Weekly Reader
Books. Weekly Reader Books offers book
clubs for children from preschool through high
school. For further information write to:
WEEKLY READER BOOKS, 4343 Equity Drive,
Columbus, Ohio 43228

This edition is published by arrangement
with Checkerboard Press.

Weekly Reader is a federally registered trademark
of Field Publications.

Copyright © 1987 by Checkerboard Press,
a division of Macmillan, Inc.
All rights reserved
Printed in the United States of America
by Checkerboard Press.
CHECKERBOARD PRESS and JUST ASK
are trademarks of Macmillan, Inc.

WEEKLY READER BOOKS presents

What Is an Iceberg?

A **Just Ask**™ Book

by Chris Arvetis
and Carole Palmer

illustrated by
Susan Swan

FIELD PUBLICATIONS
MIDDLETOWN, CT.

Sure, we'll use our globe.
I have marked the North Pole with an X.
At the opposite end is the South Pole.
It is very cold at the North and South Poles.
The snow piles up on the land.
It gets thicker and thicker.

As the glacier gets bigger and bigger, it moves forward.
The weight of all the ice and snow causes it to move.
The top layers move over the heavier bottom layers.

The glacier moves very slowly—too slowly for you to see.
As it moves, the heavy ice digs up rocks and boulders.
The glacier pushes anything in its way.

Some of the largest icebergs are found here at the South Pole.

We know of one that was miles and miles long. That's really big!

And now you know.
Icebergs are large chunks of ice that have broken off from glaciers.
The icebergs make interesting and beautiful shapes in the ocean.
But they can be dangerous, too.